RAISING THE ROOF

Other Idaho Titles of Interest
from the University of Idaho Press

History of Idaho
by Leonard J. Arrington

Nez Perce Women in Transition, 1877–1990
by Caroline James

So Incredibly Idaho!
Seven Landscapes that Define the Gem State
by Carlos Arnaldo Schwantes

One Winter in the Wilderness
by Pat Cary Peek

RAISING THE ROOF

Creating the Kibbie Dome at the University of Idaho

BY PETER T. JOHNSON

UNIVERSITY OF IDAHO PRESS
Moscow, Idaho
1998

Published by the University of Idaho Press, Moscow, Idaho 83844-1107
Printed in Hong Kong

2002 2001 2000 99 98 5 4 3 2 1
Freehand drawings used in the book are from sketches by Art Troutner.

Johnson, Peter T., 1932–
Raising the roof : creating the Kibbie Dome at the University of Idaho / by Peter T. Johnson.
p. cm.
Includes bibliographical references.
ISBN 0-89301-211-4 (alk. paper)
1. William H. Kibbie-ASUI Activity Center Dome (Moscow, Idaho)—Design and construction. 2. Roofs, Shell—Idaho—Moscow—Design and construction. 3. Wooden-frame buildings—Idaho—Moscow—Design and construction. 4. Moscow (Idaho)—Buildings, structures, etc. 5. University of Idaho—Buildings. I. Title.
TH2416.J63 1998 97-18189
690'.5827'0979682—dc21 CIP

For the creators of the Kibbie-ASUI Activity Center Dome

CONTENTS

LIST OF PHOTOGRAPHS

ART TROUTNER'S DRAWINGS

FOREWORD

BY JACK LEMLEY

The University of Idaho's unique Kibbie Dome, which won the American Society of Civil Engineer's Outstanding Structural Achievement Award for 1976, is a monument to can-do enterprise in which a dream is made reality through hard, innovative work. It demonstrates what can be accomplished within the framework of limited budgets and tight time restrictions by a team of competent, motivated engineers, manufacturers, and contractors.

It also demonstrates how a small company, Trus Joist Corporation of Boise, Idaho, using Art Troutner's innovations in wood products technology and construction, combined with a willingness to take risks, responded to the University of Idaho's request to build an indoor stadium in 1972. The proposed dome would have a 400-foot clear span, with a clearance above the playing field equivalent to a twelve-story building. Although Trus Joist was to furnish only the structural members, the company was responsible for the overall design and how it would interface with the contractors supplying the erection and support services. When Trus Joist won the bid in October 1974, it agreed to complete the project in ten months—in time for the first home football game the following fall.

This was a daunting task which required immediate, concentrated action. Unlike many of our public projects, which frequently become bogged down in the bureaucratic desire to go with a "standard" design at the expense of innovation, the construction of the Kibbie Dome is illustrative of the best that can be achieved through the design/build process. As such, the Kibbie Dome was an act of faith—faith on the part of the University of Idaho that the elegant and aesthetically innovative Trus Joist proposal could be built in an extremely short amount of time as opposed to a "proven" alternative design. It was also an act of faith on the part of Trus Joist. Even when there were significant doubts at the start of the project, the American reluctance to quit "when the going gets rough" asserted itself. It was a magnificent, innovative, aesthetically satisfying achievement, both as finished product and as an example of how the design/build process can and should work, particularly in the project's early stages.

After being awarded the bid, the engineers met only to discover they themselves had many doubts about whether or not Trus Joist could accomplish what it had proposed. Peter T. Johnson, formerly Chief Executive Officer of Trus Joist, details the events leading up to the award of the bid and the decisions that followed. In doing so, he brings to light a fascinating portrait of the challenges faced by one small company, the innovative business practices used to meet the challenges, and the true entrepreneurial spirit that energized the team to accomplish its goals. Like a barn raising on the prairie, the employees of Trus Joist came together, worked through the problems, and welded a dynamic, focused organization that got the job done and produced the largest indoor college facility in the nation at a cost of under one million dollars.

The technology the company developed based on Art Troutner's concepts for more economical ways to use forest products has helped to almost double the world's resources of lumber. Although everything was done on a first-time basis in constructing the dome, when the first two half arches came together, the tolerances were so precise that the connecting bolts could be placed by hand, a feat unheard of in wood construction. As Art Troutner points out, "It was the high point of Trus Joist's history."

Following is the step-by-step story of how a brilliant innovator and a small, dedicated organization stretched to meet the challenges and pursue the technological breakthroughs necessary to achieve success, and by doing so added a major chapter to the history of engineering, technology, and business in America in the latter half of the twentieth century.

ACKNOWLEDGMENTS

My thanks go first and foremost to my associates at Trus Joist Corporation who worked with me during the conception, design, testing, manufacture, erection, and completion of the TRUSDEK barrel arch over the stadium at the University of Idaho—Harold E. "Red" Thomas, Arthur Lowe Troutner, Zimri E. Mills, Sherm A. Nelson, James B. Lyons, William Cooley, Michael K. Thompson, Robert H. Kunesh, and Joe Smith. And my thanks extend equally to Glen E. Cline, architect of the Kibbie Dome. During our many conversations, each drew deeply from their recollections and willingly researched old files and photographs to recall the details of that memorable event in the lives of each of us. Knowing that the help of a good editor can always improve a manuscript, I turned for invaluable assistance to a world-renowned author and friend, Clay Morgan of McCall, Idaho. Terry P. Abraham, Head, Special Collections and Archives, University of Idaho Library, generously opened his records to me. I sincerely appreciate the excellent suggestions, cooperation, and faith in the project demonstrated by the staff of the University of Idaho Press. And, finally, I am grateful for the good ideas, patience, and support of my wife, Carolyn, who read draft after draft without complaint.

RAISING THE ROOF

From left to right, Harold E. "Red" Thomas, Art Troutner, and Peter T. Johnson in a recent photograph outside Art Troutner's home.

Harold E. "Red" Thomas, 71, is Chairman of the Board of TJ International, Inc. (formerly Trus Joist Corporation). Harold Thomas and Art Troutner founded the company in 1960, and Harold served as Chairman and President of the company from 1960 to 1971 and has been chairman since 1971. He earned a B.S. degree in Forestry at the University of Idaho and owned an independent wholesale lumber distribution business prior to founding TJ International. Harold Thomas currently serves on the boards of St. Luke's Regional Medical Center, Quintex Corp., Idaho Family Forum, and Clear Springs Foods of Buhl, Idaho. To relax he enjoys flying his Cessna 206 into his private ranch on the Salmon River.

Arthur Lowe Troutner, 76, inventor, designer, architect, artist, and philanthropist, was born on a farm in Pingree, Idaho, where he rode a horse a mile and a half to a rustic school. He lived with his grandmother so he could graduate from Boise High School, then attended Boise Junior College for two semesters before joining the Army Air Corps during World War II. He graduated from the University of Idaho with a degree in architecture in 1949 and co-founded TJ International (formerly Trus Joist Corporation) with Harold E. Thomas. He holds over fifty patents covering a wide range of inventions. Arthur received an Honorary Doctor of Science degree from the University of Idaho in 1976. He continues to create and spends his time between five award-winning personal residences of his design.

Peter T. Johnson, 66, received an A.B. degree from Dartmouth College in 1954 and an M.B.A. from its Amos Tuck School of Business Administration in 1955. After a two-year stint in the United States Air Force as a contracting officer, he joined a firm that constructed heavy highway and utility projects across the Northwest. He joined Trus Joist Corporation in 1968, was named President in 1971, and became Chief Executive Officer in 1975. He served as Administrator of the Bonneville Power Administration from 1981 to 1986. He is Director of the Idaho Power Company and the Standard Insurance Company of Portland, Oregon. Peter lives in McCall, Idaho, where he sculpts, paints, and writes.

RAISING THE ROOF

GREAT ARCHITECTURE is the visual music of a civilization, performed in three dimensions and resounding through time. It is a visible echo of who we are at our best and most creative. For over 5,000 years, monumental constructions have been the hallmarks of the spirit and vigor of human societies. Stonehenge, Egyptian pyramids, Greek columns, the Inca Temple of the Sun, China's Great Wall, Roman aqueducts, the medieval cathedrals—all have sung of the aspirations and triumphs of their eras. Some still stand, and they sing to us today of their beauty, cooperative enterprise, and personal genius.

American architecture in the last 150 years has produced stunning examples of form and function: gracefully strung bridges; gigantic yet elegant dams; and buildings so tall that people call them skyscrapers both for their appearance and for their daring. In this same tradition a new kind of structure—also exemplary in aesthetics and function—rose into the sky on the beautiful Palouse Country of northern Idaho in 1975. Like the capacious dome of the Roman Pantheon and the graceful flying buttresses of Chartres Cathedral, the Kibbie-ASUI Activity Center Dome seemingly defied gravity. Its first ever, first anywhere, 400-foot clear-span wooden trussed-barrel arch offered the community a weather-free forum for athletics, education, and the arts.

The structure's simplicity, harmony, strength, and grace resulted from the sweat, pluck, and genius of a team of committed individuals, and it was built at the remarkable cost of less than $1 million. Its dreaming, its design, and its dimensional elegance reflected the spirit of a company and a university, and it revealed the character of the individuals who dared build a vision.

This facility began as a dream eight or nine years ago. . . . Its completion has surpassed our fondest expectations. It houses and provides space for every outdoor sports activity that we sponsor at the University of Idaho. . . . Most of all, it focuses attention on the University of Idaho and provides us the opportunity to schedule special attractions that beckon cordially to all.

—Leon G. Green, Director of Athletics

Prudence or Audacity—A Vision Emerges

EARLY IN 1972, a young Idaho company was just stepping over the threshold into its future. The Trus Joist Corporation of Boise, Idaho, had already earned a notable reputation among North American architects, engineers, and builders, and the steps it would take over the next three years would bring it lasting fame.

Trus Joist's revolutionary engineered wood structural products utilized less wood fiber from the world's valuable and vanishing forests and supported heavier loads than traditional dimensional lumber, allowing for greater design flexibility. These products had grown out of the genius of a University of Idaho graduate in architecture—Arthur Lowe Troutner.

Troutner is an artist of ideas; his brainstorms flash with high art. Like Thomas Edison and Buckminster Fuller, his talents are many-hued and multidirectional. The old saying about inventing a new mousetrap rings more than true for Troutner. Including his famed wood-fiber enhancing designs, Troutner holds over fifty patents. Examples of his inventions range from racing sailboat hulls, lighter-than-air dirigibles, and vertical access wind turbines to solar mirrors and even a rapid-fire rubber-band gun.

Troutner founded Trus Joist in 1960 with another University of Idaho graduate, Harold E. "Red" Thomas. Thomas has his own style of genius as an entrepreneur and marketing visionary. He could see potential in Troutner's designs beyond what Troutner himself saw, and together they gathered a team of talented engineers, savvy managers, and technical sales people to nurture and develop Art Troutner's brainchildren.

Troutner and Trus Joist engineers progressively developed a complimentary family of proprietary open-web trusses which used machine stress-graded wood chords, galvanized steel tubing, and pin connections.

I frequently heard architects and engineers refer to the company's products as "all lean meat." Trusses for specific applications, ranging from residential through commercial to light industrial structures, with clear spans of up to 100

From left to right: Two revolutionary TJIs (using MICRO=LAM® flanges and 3/8-inch plywood webs) which are replacing solid sawn floor joists in residential construction. Standing alongside are MICRO=LAM® engineered wood products in varying depths for applications such as headers and scaffold planks.

Right: Trus Joist's popular, up to 40-foot span, TJL open web series joist, which utilizes machine stress-rated 2 by 4s as top and bottom chords. Left: TJ 50 joist for longer spans which utilizes double top and bottom MICRO=LAM® chords. Similar joists, the TJM and the TJH are the vertical structural members used in the Kibbie Dome's end walls.

feet, were custom designed on a computer to harness the greatest contribution from the intrinsic properties of each material.

The company's growing expertise was soon tested. In 1972, Glen Cline, a distinguished Boise architect, was selected as the architect-of-record for the University of Idaho in its quest to build an indoor stadium protected from the extreme weather of Idaho's panhandle. Cline had utilized Trus Joist products for years and was especially impressed by the flexibility of the company's technologies.

But Glen Cline was drawn by more than Trus Joist's reputation. During the Great Depression, while Cline was working toward a degree in architecture, France and Brazil had indirectly helped to develop his predisposition for the freedom and flexibility that Trus Joist technologies now afforded. Cline had put himself through Kansas State University's architectural school partly by working in the architectural library where, during free moments, he devoured books and magazine accounts of acclaimed architects and their achievements.

"I was fascinated with Le Corbusier," Cline said of the famed Swiss architect, whose innovative works could be seen in France and Brazil. Le Corbusier was a pioneer architect who ceased using traditional forms, such as arches, and began designing buildings that were expressions of their function. Good illustrations of Le Corbusier's innovations are the Swiss and Brazilian students' pavilions

Glen E. Cline, 78, architect and planner, was born in Fredonia, Kansas, in 1920. He received his B.S. degree in Architecture in 1949 from Kansas State University. He is a registered architect in Idaho, Utah, Montana, Oregon, Washington, and California and was principal architect of Cline, Smull, Hamill & Quintieri Associates from 1969 to 1985. He served as Chairman of the Idaho State Board of Architect Examiners between 1956 and 1969. He was the principal architect and project manager of the Kibbie Dome, the Pavilion at Boise State University, and the Boise Air Terminal. He was honored as a Fellow by the American Institute of Architects in 1985.

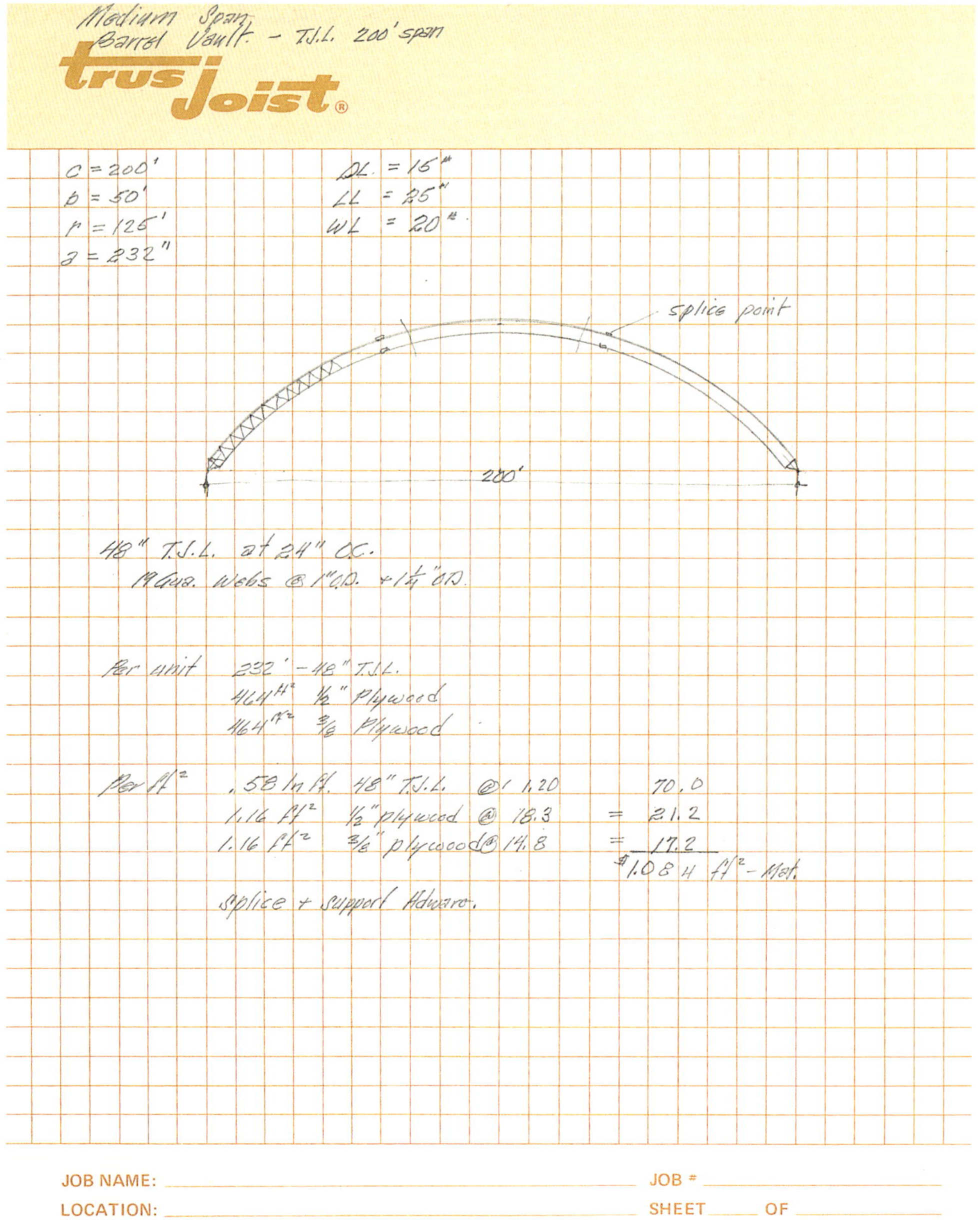

Art Troutner's drawing and calculations of a prospective application of the TJL open web truss.

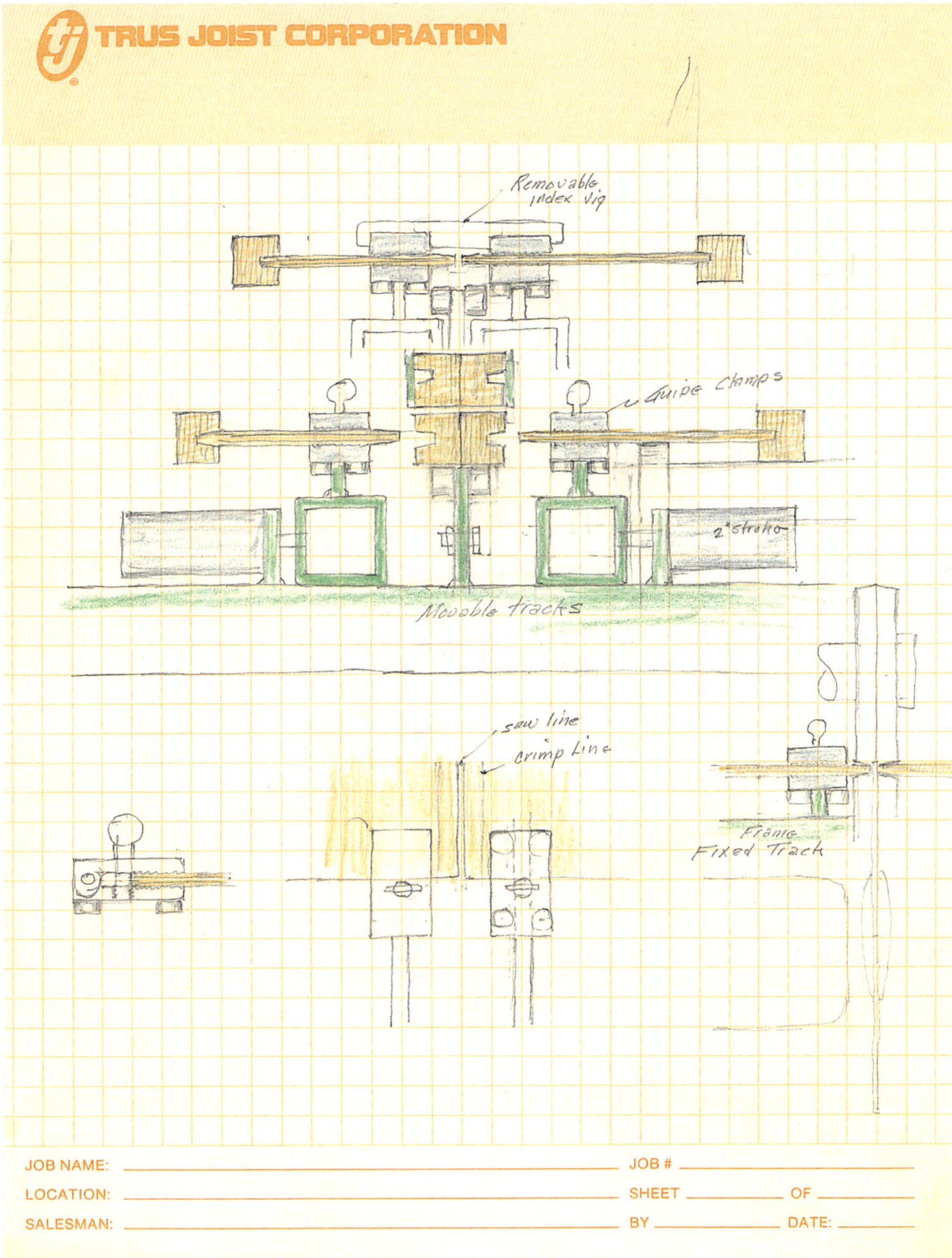

Freehand machine drawing by Art Troutner.

Sherman A. Nelson P.E., 64, received his B.S. degree in Civil Engineering in 1957 and his M.S. degree in Civil Engineering in 1962 from the University of Iowa. Sherm, a Registered Professional Engineer in Idaho, Oregon, California, and Washington, served as Trus Joist's Chief Operations Engineer from 1965, then Chief Engineer from 1980, and concluded his career with Trus Joist in 1996 as Vice President of Technology and Engineering. Sherm has served as President of the Idaho Section of the American Society of Civil Engineers, the Idaho Society of Professional Engineers, and the Structural Engineers Association of Idaho. He is currently a Consulting Engineer in Oceanside, Oregon.

at Cite Universitaire in Paris. "Le Corbusier was the Trus Joist innovator of the '20s and '30s. When Trus Joist came out with its first products I saw infinite possibilities. With the company's technical help I gained freedom to create designs that pleased my clients, much as Le Corbusier had done. Everything Trus Joist did tested out and worked fine. And everything Art touched worked like magic."

James B. Lyons, 74, was born in Salt Creek, Wyoming. Jim attended public school in Billings, Montana. After serving two years during World War II in the Navy Air Corps, he graduated from Montana State University in 1948 with a B.S. degree in Industrial Engineering. Jim excelled as an athlete in football, track, gymnastics, tennis, skiing, and golf. He was a general contractor for ten years and then a marketer of glulam beams before joining Trus Joist in 1963, where he managed marketing and was Director of New Product Development. Jim retired in 1980 to travel, golf, and enjoy his grandchildren.

As Chief Executive Officer at Trus Joist, I realized that Cline and Troutner had more than magic in common. Besides being an inventor, Troutner was also an architect. While Cline was designing his commercial structures in the Pacific Northwest, Troutner was designing homes in Sun Valley, Idaho, which set new standards for residential beauty.

Creation is beautiful and Troutner was creative in everything he approached. He sought optimums in the utilization, manufacture, and final assembly of materials. Solutions often came to Troutner in midstride, after he had picked up his chalk to illustrate an idea on a blackboard. When he started drawing, I couldn't keep my eyes off his hands. When Troutner picked up a pencil or chalk and drew, it was like opening a window into his mind. Troutner was also ambidextrous. As colleague Sherm Nelson recalled, "At the blackboard, he'll draw with his right hand and write with his left. Then he'll write with his right and draw with his left!"

Once Troutner began to illustrate an idea it would invariably evolve, then change more, arousing excitement in anyone watching. His steady hand drew straight lines and perfect circles revealing one astonishing idea after another.

A salesman told me that an architect in Texas once drove by a commercial construction that employed Troutner's structural wonders. The architect slammed on his brakes and nearly caused a major accident. "I had to stop," he said. "What I saw was beautiful."

Few obstacles withstood Art Troutner's vast imagination and intelligence. But like many geniuses, Art could be quite independent and extremely stubborn on occasion.

The University of Idaho was, as usual, strapped for funds and Cline had to be creative in his approach. So in the spring of 1972, he took his idea about the stadium to the Trus Joist offices, located in a copse of locust trees just below

Boise's West Bench. There he found Art Troutner in his office, which was a confusing and cluttered opposite of Troutner's graceful and simple designs. Magazines, papers, scale models, product samples, and books covered every available surface, except for a small clearing in the center of the desk. That clearing, like Troutner's mind, was always ready for another idea.

"Art," Cline asked, "can you adapt any of your technologies to cover a 400-foot clear span and give me a clearance equivalent to a 12-story building for the playing field?"

At that moment, Troutner's hands settled upon a clean sheet of 8½ by 11 inch graph paper in the clearing.

Sherm Nelson, our Chief Operations Engineer, enjoyed a reputation as an engineer's engineer. When Nelson first learned from Zimri Mills, Trus Joist's Chief Engineer, of the possibility of the company designing a cover over the athletic stadium at the University of Idaho, he remembered what his distinguished father-in-law, who was also an engineer, had said, "If a structure can be sketched on an 8½ by 11 inch piece a paper, it can be built."

While Art's imagination was exploring a number of concepts, the arch configuration progressively appeared to be the superior choice. But Art was not the ultimate expert in the more mundane requirements of rational stress analyses essential to answering such questions as structure height and radius. Sherm knew where he could locate a published, creditable protocol to assist Art—the *Timber Design Handbook*. He gave it to Kevin O'Sullivan, Troutner's young and precocious research and development engineer, who programmed the procedure into the company's computer.

Nelson had great confidence in O'Sullivan's structural sense and analytical abilities, but Art Troutner's respect for O'Sullivan was critical because he didn't trust all engineers.

O'Sullivan had partly earned Troutner's respect because, when the Trus Joist team played their late night poker games, O'Sullivan would remove his artificial glass eye and place it on a shelf directly behind Troutner so it looked over his shoulder. The inventor denied that this made him nervous, but O'Sullivan won often, and he attributed his winnings to his "visionary" knowledge of the cards.

Zimri E. Mills, 73, was born in Wilder, Idaho, and graduated from Wilder High School in 1943. He went directly into the U.S. Army, served for two years under General George Patton, and fought in the Battle of the Bulge. He entered the University of Idaho in 1947 and graduated with a B.S. degree in Agriculture Engineering in 1950. He joined Trus Joist Corporation in 1963 and served as Chief Engineer until his retirement in 1980, although he has continued consulting for the company. He is less retired than pursuing new careers as a world traveler, as a designer and constructor of a summer cabin on Cascade Reservoir in Idaho, as a developer of residential subdivisions, and as a farmer.

> *In less than one year, the William H. Kibbie-ASUI Activity Center has brought local, state, and nationwide recognition to the University of Idaho campus. The "Dome" is a visible symbol of progress at the university and has become a point of interest to our alumni and friends throughout the country. . . . In addition, the center has already assisted us greatly in recruiting athletes and other students to the University of Idaho.*
> *—Dick Johnston, Director of Alumni Relations*

Troutner and O'Sullivan began working with James Lyons, the Director of New Product Development, and ideas quickly emerged including geometric variations on the barrel arch and a rectangular space frame configuration. As Trus Joist's top managers, Red Thomas and I were aware of this prospecting work but we hadn't paid much attention. Such projects rarely happened, particularly ones of this magnitude. Besides, at the time the company was midstream in the introduction and marketing of two of its breakthrough engineered wood technologies, both designed by Art Troutner. MICRO=LAM®, a laminated veneer lumber, would become a way for trees to be converted to a higher value. The other technology, the TJI® all-wood I-beam, would go on to replace 2 by 10s as the preferred floor joist in residential construction. This all-wood I-beam used MICRO=LAM® for its top and bottom flanges. As an example of Art's environmentally sensitive genius, the highly efficient TJI® required about one half the wood fibre to perform the same structural function as a solid sawn 2 x 10 which it replaced. It is fair to hold that by his inventions Art Troutner roughly doubled a significant portion of the world's forest resources.

Jim Lyons was an engineer by training, with years of experience in the building construction industry. He could also work effectively with Art's limitless imagination and the more prosaic design approaches of professional engineers. Lyons had worked with Troutner and Thomas in the development of the company's open-web trusses, and he understood the extraordinary challenges of converting new ideas and technologies into commercial realities. Harmonizing these various perspectives would be crucial to the stadium project.

The Trus Joist TRUSDEK roof system (TRUSDEK is a roof or floor structural system using wood top and bottom chords restrained apart and connected to steel tubing in triangular patterns) would further complicate the relationships between the participants. Trus Joist was to supply only the structural materials for the barrel arch and subcontractors would supply erection services and support systems, such as the acoustical ceiling, mechanical ducts, sound system, lighting catwalks, dividing nets, and even the scoreboard. However, Trus Joist was responsible for the overall design and interfaces with the subcontractors.

Lyons received invaluable assistance from Zimri Mills, Trus Joist's Chief Engineer. Blessed with intuitive good sense, Mills was always open to new ideas and understood the testing required to prove a new product. Because of the unique nature of Trus Joist's original structural products, equally imaginative tests had to be conceived that would prove to the company, professional engineers, and code officials that the products would perform to code requirements. The TRUSDEK barrel arch (a TRUSDEK barrel arch is a roof system but a TRUSDEK roof system is not necessarily a TRUSDEK barrel arch; it could be a rectangular shape) would be no exception.

> *If a man make a better mousetrap, the world will make a beaten path to his door. —Ralph Waldo Emerson*

Meanwhile, Troutner and O'Sullivan were letting their imaginations run free. One idea they came up with was to construct a MICRO=LAM® press right on the playing field. The press would extrude billets (slabs of laminated veneer lumber 2 feet wide by about 2 inches thick) in a continuous 455-foot ribbon directly into the roof system. The press would then be shifted two feet to the side and the procedure repeated for the next roof support member.

Troutner liked this idea because it would eliminate the need for any splices within the span. However, the scheme had major problems, not the least of which was that in 1972 the final technology and design for a MICRO=LAM® continuous press had not been proven. Furthermore, the adhesives, laminations, and

Artist's preconstruction vision of athletic activity within the Kibbie-ASUI Activity Center Dome.

densifications for the MICRO=LAM® products themselves had yet to be finalized. We were only in the early planning stages of opening a large MICRO=LAM® manufacturing plant somewhere in the Pacific Northwest, and that would happen only if the technology proved feasible. All that existed in Boise's new products plant was an experimental prototype press, the first of its kind anywhere in the world. Most of the time it was shut down for modifications.

Many other design components of the TRUSDEK structure were original concepts, and each had to be individually tested at full scale before it could be relied upon. The design loads for a structure this large could not be picked from a handbook. Still, Troutner decided that a trussed-barrel arch held the greatest promise to meet the stadium cover challenge, even considering all its unknowns.

Consequently, Zimri Mills and Sherm Nelson recommended to me that Trus Joist hire an independent, highly qualified engineering consulting firm to verify the final design and lend the concept more credibility, since Trus Joist's reputation was in specialty engineering as a product supplier. Previously, Nelson had worked on three-dimensional, "space frame" structural projects with KKBNA, an esteemed structural engineering firm in Denver, Colorado. KKBNA was hired and its representatives immediately came to Boise to meet with Trus Joist officials,

The addition of the Kibbie-ASUI Activity Center to the University of Idaho campus will enable students to participate in "outdoor sports" all during the year, as well as improve greatly our ability to accommodate the public at large University events. . . . With the addition of this expanded all-weather playing area, nonvarsity student participation in many sports has increased noticeably. The "Dome" has already demonstrated its versatility in accommodating events such as drill team practice, a local Chamber of Commerce Home Show, several high school football games, and semester registration, to name a few. Currently there is a "wheelchair olympics" event taking place which would not have been possible a year ago. . . . In short, the Kibbie-ASUI Activity Center is everything we anticipated and more. Because it is so new, we are well-aware that we have only "scratched the surface" when it comes to potential uses, and therefore look to the future, confident that our decision to construct a multiuse facility of this type was a wise one that will produce future dividends.
—Sherman Carter, Financial Vice President

Glen Cline, and Cline's civil engineer from CH2M Hill, then a Boise-based engineering firm.

Between October 1974 and February 1975, KKBNA's Jerry York and Trus Joist's key players were busy. Copious drawings and computer runs confirmed the basic design and established preliminary loads. Now a good construction contractor had to be found.

Earlier in 1974, I had introduced Jim Lyons to a longtime business associate, Tom MacGregor, President of the MacGregor Triangle Company. MacGregor had extensive construction experience, including the erection of high-voltage transmission lines under difficult and challenging conditions, including construction of one of the first helicopter-erected lines from Hells Canyon Dam on the Snake River up the shear rock escarpments on the Oregon side of the deepest canyon in North America. He had expressed an interest in submitting a proposal to assemble and erect the structure in Moscow, Idaho. Lyons and MacGregor worked out an understanding after spending hours discussing the staging areas, preassembly, crane sizes, and the special rigging that would be required to avoid creating excessive stresses on the preassembled arch sections during erection.

The consortium still needed a prime contractor, primarily to manage the many subcontractors. Tom MacGregor recommended Emerick Construction Company of Portland, Oregon, which agreed to act in that role.

It fell to Jim Lyons to coordinate this team to prepare a preliminary design and bid the entire roof system. Lyons recalled, "Glen Cline, KKBNA, MacGregor, Emerick, and most of the major subs worked hard to stay on top of their responsibilities. The project was truly amazing when considered from any viewpoint. It was unique the way the project was conceived, designed, bid, managed, and constructed."

About a year before the project was let for bid, Lyons prepared a preliminary estimate of material and labor costs for the MICRO=LAM® TRUSDEK barrel arch design. "These were ball park figures," Lyons said, "to help guide Glen Cline and the Regents of the University of Idaho in determining their funding needs. There is no doubt Glen liked Trus Joist and our concept because he gave us every encouragement up to the bidding process. Glen's confidence in us

William Cooley, 77, a third-generation Oregonian, was born near Portland, Oregon. Having just graduated from high school in 1939, Bill joined the Oregon National Guard, where he was mobilized into active duty and served with great distinction for five years until the end of World War II. He enrolled at the University of Oregon and received a B.S. degree in Architecture in 1951. Bill worked in the plywood industry until joining Trus Joist Corporation in 1971 as General Manager of the MICRO=LAM® Plant in Eugene, Oregon. After retiring from the company in 1987, he has improved his golf handicap between travels about the world.

Joe Smith, 65, was born in Nampa, Idaho, grew up on a farm south of Nampa, and graduated from Nampa High School in 1951. Joe served two years in the U.S. Army before joining Trus Joist Corporation in 1961. He set up and launched early Trus Joist manufacturing plants in Portland, Oregon, Phoenix, Arizona, Santa Rosa, California, Dubuque, Iowa, and Claresholm, Alberta, Canada. Next he was Manager of the New Products Plant, which teamed Joe with Art Troutner. During Joe's tenure, they set up the machinery and test marketed many products including the TJI® all-wood I-beam and MICRO=LAM® laminated veneer lumber. Joe retired from Trus Joist in 1987 and now raises fruit and cattle on a farm near Nampa.

and his fair handling of the many problems prior to the bidding helped convince us to hang in there and see it through."

Lyons continued, "It was after we submitted our idea that he [Glen] decided on the difficult route of design-construct. That was about the only way the law would allow Trus Joist's proprietary system to be bid into the job. The design-construct approach is usually employed on smaller projects. This must have been one of the most complex applications of this bidding concept ever. In addition to the difficult job of writing the performance specifications, the architect had to schedule several prebid conferences to explain to all bidders what would be required. Traditionally, building contractors and their subs and suppliers are geared to pricing and bidding from a complete set of plans and specifications, which they can do quickly and at little cost. The design-construct approach carried many uncertainties and risks. The Trus Joist proposal would be particularly complex."

Pivotal Events

I REMEMBER THE day in 1974 when the University opened the sealed bids in Moscow, Idaho. That morning, having experienced many times the bid, award, construct cycle, I felt the customary anxiety that prevails between the offer and acceptance phases, as one might await response to a marriage proposal. Our offer could be rejected for any number of reasons or we could get the job and surrender a portion of our lives to constructing it. Little did I suspect the marriage that would unfold. The only other bid was a glulam trussed arch design, but it was considered a clumsy, incomplete, and nonresponsive proposal. So Cline announced that, time being of the essence, he would proceed expeditiously to finalize a contract with Trus Joist using its TRUSDEK proposal.

Trus Joist and its design-bid team would be on a fixed-bid, all-risk contract, including stiff monetary penalties, called "liquidated damages," that would be levied if the facility weren't ready for the first football game ten months away. These damages, meant to offset the University's revenue losses, could quickly compound into millions of dollars. Once the contract was signed, either a fully-completed barrel arch would be in place or Trus Joist might face bankruptcy.

In our many discussions with Glen Cline leading up to the bid there had been no mention of liquidated damages, though such conditions were commonplace in contracts of this size and nature. Now the contract forcefully spelled out the penalties. No longer would the company have the luxury of dreaming up neat and nifty schemes. It might soon be in a make-or-break business deal.

Glen Cline had just presented us with the company's biggest challenge in its brief, shining history. Now Trus Joist had to take a last close look at the project before jumping in with both feet.

Up to this point, I had only been monitoring the project's possibilities. Now it was within days of becoming a business reality and it demanded a vigorous reward/risk analysis. I knew construction well, having cut my teeth in the business. Thirteen years with a large highway and utility contractor had taught me the pain of failing to properly organize a project, as well as the great benefits that come from a well-managed undertaking. The gap between these two extremes is vast. Nothing could be left to chance. The construction business has enough uncertainties without knowingly introducing more.

Now I wanted to understand firsthand the current status of the design: who would be responsible for what, which suppliers had been retained, and what the unresolved issues needing decision were. Further, the contract's severe liquidated damage provisions had introduced a new urgency.

Fiscal year 1974 was about to conclude, with Trus Joist recording sales of $42 million. Should Trus Joist take on a $933,000 all-risk contract to finalize the design, manufacture and supply of a 400-foot clear-span, first-of-a-kind barrel arch over the football stadium at the University of Idaho? An award had to be made right away.

Realizing that I had to finalize the schedule, implement the accountabilities and organization, and set priorities against other company endeavors, I called a meeting of the project's key employees. Most of the members of top management were present. In the Trus Joist conference room their voices betrayed an excitement subdued only by a realization of the immensity of the undertaking.

The meeting began with reports from Jim Lyons, Zimri Mills, and Sherm Nelson, as well as from Robert Kunesh, the company's highly regarded

Robert H. Kunesh, 68, was born in Chicago, Illinois. He attended the Illinois Institute of Technology and the Universities of Illinois, Michigan, North Carolina State, and Oregon State, majoring in engineering, forestry, and wood technology. He earned a B.S. degree in Forestry and a B.S. and M.S. degree in Wood Technology from the University of Michigan. Bob worked in wood technology at the Weyerhaeuser Company, researched the strength properties of wood at Oregon State University's Forest Products Laboratory, and served as Director of Wood Technology at Trus Joist Corporation. Currently he consults in the field of laminated wood products.

wood technologist, and Bill Cooley, General Manager of the new MICRO=LAM® plant, which was in the early debugging phase of commercialization in Eugene, Oregon. The meeting started well, but soon digressed into an alarming recitation of Art Troutner's doubts. Like an adversary, Troutner challenged each presenter with critical questions on details. It was almost as if he had not been part of the process up to that point. I listened carefully, my anxiety rising, as Troutner challenged one unresolved issue after another. For example, the design was not finalized and there were no final decisions yet on billet sizes or on web sizes. No final decisions had been made on plate connections or on the clips that would connect webs to billets. Panel point distances remained unspecified, as did the structural connections of the TRUSDEK to the bearing foundation and the vertical end-enclosure structure. Then there were erection procedures and related matters to consider. The company had little more than the concept and the preliminary analyses indicating that it was feasible.

Joe Smith, Manager of Trus Joist's New Products Plant and soon to become a key player in the project, recalls, "I learned about the project after we got the bid. Wasn't anything designed! Everyone was there at the first big meeting. While discussing all the work to be done the mood turned nervous. It really became tense when the subject of the billets taking on moisture came up."

Bob Kunesh had performed preliminary studies on the expansion properties of MICRO=LAM® billets when they were exposed to moisture. Kunesh had taken a 2 by 3-foot billet and placed it in a tank of water. The billet had swelled across its width by 1¼ to 1½ inches. His conclusion: If every billet in the 192-billet-wide barrel arch became soaked, the arch would overhang its foundation by between 20 and 24 feet.

I asked what would be the result if that occurred.

Joe Smith answered, "There's no way in the world you can put them up."

This revelation scared everyone. Art Troutner exploded, forced his chair back, stood, and moved toward the blackboard. I too jumped up and began pacing the room with my hands tightly clasped behind my back. Joe Smith told me later that I swore a blue streak and looked like a panther stalking its prey.

The meeting was serving a single frightening purpose: to illustrate how little had been done and how very much remained to be accomplished. I thought of the company's two world-class research and development projects currently being developed: the massive MICRO=LAM® manufacturing commitment in Eugene and the promising new TJI® all-wood I-joist. These were usurping the time of everyone in the room. I also thought about the liquidated damages for non-completion of the stadium cover. And then I abruptly adjourned the meeting.

I closed myself in my office and tried to assess what I had heard, to put the stadium project into context, to estimate the people and time required to complete the project successfully. I considered the risks of failure to the company financially and to its reputation. I recalled the enthusiasm and effort key personnel had invested in the project. I got up from my chair and looked out the office window, rehearsing in my mind the project's possible benefits and very real risks. High above Boise I could see the Bogus Basin Ski Area, where Trus Joist MICRO=LAM® billets had already been used for practical and artful effects. During the debugging of the first MICRO=LAM® press in Eugene, Oregon, a large quantity of billets had been manufactured that could not be certified for optimal use. Coincidentally, the contractor building a new lodge at the ski area had encountered a cost hurdle and couldn't afford the specified tongue and groove 2 x 6 flooring and roofing material. Zim Mills approved a nail-laminated system of bargain-priced, ripped MICRO=LAM® on edge which met the contractor's need, found a market for Trus Joist's dead inventory, and created an unexpected aesthetic effect with the thousands of ⅛-inch MICRO=LAM® veneers. If only the University of Idaho stadium project didn't have so many unknowns . . .

When I got home that evening, I told my wife Carolyn that I felt I was trapped. I blamed myself for not keeping closer tabs on the project, and I wondered whether the others understood the serious fix the company was in.

As I mulled the situation over, one impression from the morning's meeting kept returning: Art Troutner's pyrotechnics and apparent sudden lack of personal commitment to the project. Perhaps that was the way of an eccentric genius, but there was no room for games now. The project needed Troutner's unequivocal commitment and attention to detail if the company was going to succeed. If Troutner stood apart, the project would fail.

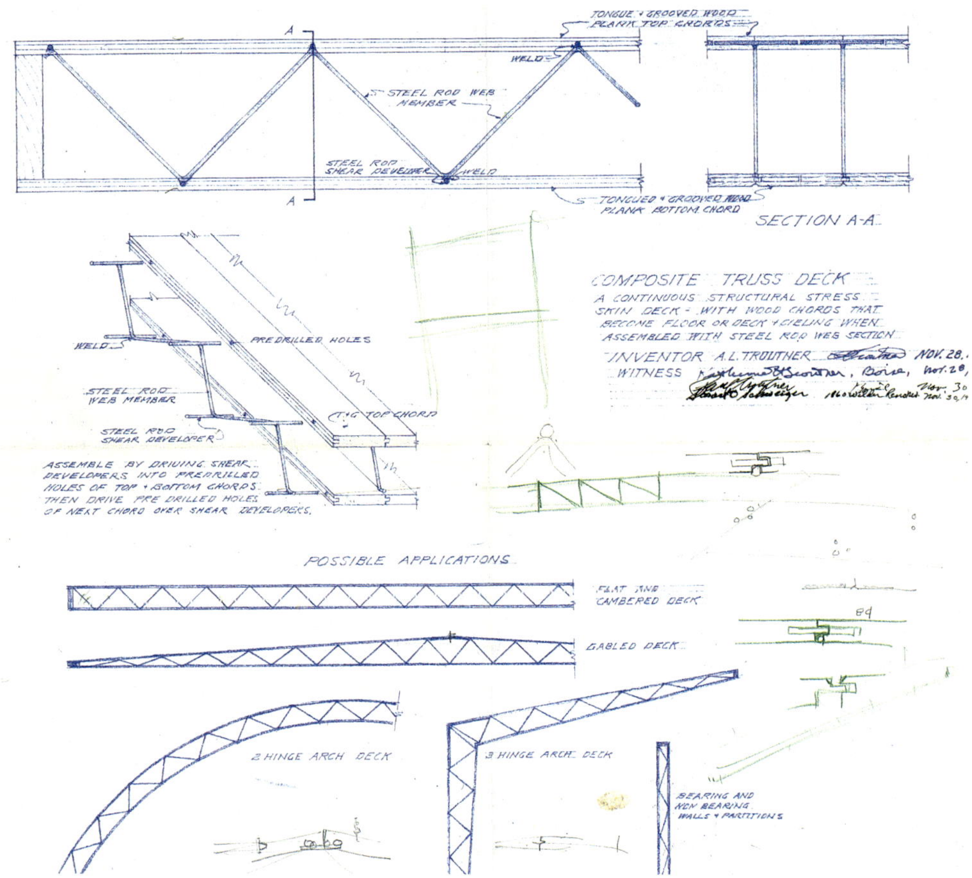

Sketches by Art Troutner in support of a patent application.

By ten o'clock that evening, I had concluded that I had to confront Troutner. If Troutner vacillated, the best business decision was to forfeit the $100,000 bid bond and suffer the ignominy of backing away from an exciting project for which the University of Idaho and the architect and my good friend, Glen Cline, had been led to believe the company was ready to perform.

But forfeiture was better than crippling the company upon which employees, customers, and stockholders relied. Trus Joist could recover from an embarrassment, but probably not from a financial hemorrhage and a new reputation for nonperformance. I picked up the phone and called Troutner.

"Art, we have a potential disaster on our hands. Today's meeting made me ill. I'm convinced that the design and specifications for the stadium cover have not been advanced adequately to allow our company to assume the associated risks of the contract. Neither you nor anyone else had satisfactory answers today to a multitude of questions. All we have is a general concept.

"Even worse," I continued, "you were nitpicking everything on the project. You've checked out. And I know you are the only person who can lead us to the best decisions.

"Only one alternative is left. I'm forfeiting our bid bond tomorrow morning. I have no other choice than to back away from the project."

At first there was silence on the line. Then Troutner blurted angrily, "Peter, how can you even consider such a thing?"

I remained firm and terse on the phone. The time for debate and friendly discussion had passed. I told Art I would call Glen Cline in the morning. The phone conversation ended.

At 4 A.M. it was my phone that rang awakening me from a fitful sleep.

"Peter, this is Art." Troutner wouldn't be interrupted. "You can't back out. Neither you nor I will have a job if you do. There aren't that many problems that remain. None are insurmountable. Haven't slept. Been thinking all night."

Troutner ran out of breath long enough for me to say, "Art, this is the first time I've heard you talking about solving the challenges that remain. But I have no confidence you will follow through. Tomorrow morning you'll forget this and go off on another idea."

Art interrupted, his voice resolute. "Peter, you've misunderstood me. But now you've gotten my attention. I'm fully behind the project. It is an exceptional application of our technology, which has already been proven on smaller projects. I will devote top priority to it. I understand the risks you envision, but I also know the concept will work. It will require a lot of effort and finding the best answers won't be easy, but my commitment to the project is unqualified."

"Art," I said, "the only way I might reconsider my decision is if you make a firm commitment. Time's run out for another conversation like this. Your pledge must be unwavering."

"You've got it," Troutner responded without hesitation.

When Art Troutner entered his office at 7:30 A.M. that morning, I was there to meet him.

Michael K. Thompson, 53, was born in 1945 in Lethbridge, Alberta, Canada, where he received his high school education. He earned his B.A. degree in Languages in 1969 and an M.B.A. degree in Business Administration in 1971 from Brigham Young University. After one year with Xerox Corporation, Mike joined Trus Joist in 1972 as a management trainee. He has been the Plant Manager of three major Trus Joist manufacturing plants from Boise to Valdosta, Georgia, Project Manager of the Kibbie Dome project, and General Manager of Canada and three regions of the United States. Currently Mike is the General Manager of the company's operations in Japan.

"I will match your commitment, Art, with one of my own. I'm going to devote the time to ensure we manage all activities in an effective manner. I've been thinking about how to organize the project. I will be implementing that plan beginning this morning. Together with the commitments of everyone else in the company, we will maximize our prospects of succeeding. With your promise, Art, I will not throw in the towel. But if the Board of Directors knew the risks the company is assuming in this project, they would fire me."

We shook hands and went to work. The contract was executed and I submerged myself into the organization and management of the project.

Concept to Design

BY ACTING QUICKLY I was making my personal commitment clear and demonstrating the priority and urgency the project would enjoy within the company. I formed a steering committee of the key talents who would be required to complete the project on time, a management technique that had proven highly effective in the commercialization of MICRO=LAM® and the TJI®. The Steering Committee included Art Troutner, Zimri Mills, Jim Lyons, Sherm Nelson, Bill Cooley, Joe Smith, Bob Kunesh, and a young talent in the company, Michael Thompson. This group began meeting the day the contract was signed and continued meeting until all requirements of the project were successfully performed. The meetings, in Boise and Eugene, always concluded with written action assignments and deadlines for the responsible parties.

I believed that the project needed a project manager, someone to ensure daily coordination, communication, and performance. Harold Thomas and I agreed on Mike Thompson. Thompson had been doing an outstanding job managing the company's open-web manufacturing plant in Boise.

It was Thanksgiving Day. Mike Thompson's wife Carol had finally assembled her entire family for dinner for the first time ever. Fifteen relatives were just sitting down to the table when the telephone rang.

"Mike," I said, "sorry to interrupt your evening, but could you meet me for breakfast at the Rodeway Inn tomorrow at 7 A.M.?"

"Yes, of course." Thompson had no idea what I had in mind.

Thompson was there early, a habit that would serve him well in the assignment I was about to offer.

"Mike, you know about this stadium cover project we have been awarded. It's going to be a challenge exceeding anything our little company has faced before. I've concluded that the project needs an overall project manager. Right now, everyone else involved is focused on his specific task. That leaves a lot of loose ends."

I went on to describe in general terms the contract. I concluded with the penalties Trus Joist would suffer if the project was not completed on time.

"Mike, it will be a tough assignment but a great learning experience. I want you to start immediately. Will you accept?"

"Yes, I sure will," Thompson said without hesitation. "I'll work myself out of the Boise plant as fast as I can. Then I'll get with Jim Lyons to get up to speed on where we are." Thus began the assignment that would consume over a year of Mike Thompson's life.

One of six 30-foot long, mock-up sections of TRUSDEK in jig. Load cells mechanically pull cables until section destructs to determine and confirm engineering values and component designs.

> *Everyone involved in that project felt he was part of something new, something unique, something he could be proud of.*
>
> *—Glen Cline, Innovative Architect*

Thompson spent the next three weeks working with Jim Lyons. He then left Boise during the Christmas holidays on a planned trip to spend a few days with his family in Canada. When he returned he found a note on his desk.

"We have a three-ring circus going on and it needs a ringmaster—get busy!" It was signed Jim Lyons.

Even though he and Carol were expecting their first child in June, Thompson never let up on his oversight of the work, which spread across three states. Nor did he neglect Carol. Because flights were so unpredictable from Boise to Lewiston and Pullman, it was not uncommon for Mike to leave Boise in his green Volkswagen Bug in the early hours of the morning, drive 350 miles to Moscow, arrive before noon, perform his work, then turn around and drive back to Boise, arriving after midnight.

To confirm the project's urgency, I flew on December 3, 1974, to Eugene with Thompson and Lyons to meet with Bill Cooley, General Manager of the Eugene MICRO=LAM® plant, and his staff. Thompson's role was explained and initial activities and accountabilities were assigned.

A month and a half later, as the design was being finalized, I realized that some activities needed the authority of my office, so I assumed an even more active role in the project. At the Steering Committee meeting of January 26, 1975, I announced, "Because of the complexity of our Moscow stadium cover project and the importance of timely decisions between the several parts of our company, I am moving into an active position with Mike Thompson to assist him at this time in key problem solving decisions. This in no way alters his assignment or anyone's previous relationship."

Three memoranda encompassed thirty critical actions dealing with billet manufacture, the precise details and accountabilities for the pin connections in the clips, and Bill Cooley's time schedule specifying the exact dates for billet shipments from Eugene.

One of Mike Thompson's early assignments was to work with Jim Lyons to prepare a comprehensive, time-scaled Critical Path Network Analysis outlining all activities and events as well as their order in time. I had gained great respect for this scheduling method while working on heavy utility and highway projects with the U.S. Army Corps of Engineers. It allowed the Steering Committee to see the entire project laid out in detail on one sheet of paper, albeit a large one.

"Peter insisted that a network analysis be prepared for the stadium cover," Thompson recalled. "Although I hadn't worked with Critical Path Scheduling since college, Jim Lyons and I went into the conference room and began preparing one on the blackboard. Whenever we thought we had all the activities and events in order, we thought of something else and had to erase and start over. Peter came in from time to time to warn us of 'lies and falsehoods,' places where we were not being realistic and fully honest with ourselves. It took us a long time, but the final network brought into sharp focus the daunting work and unforgiving schedule before us. We would have been lost without this management tool."

A bonus of this technique was the cohesion, empathy, and motivation it built within the Steering Committee. Everyone knew exactly what was expected of him as well as of everyone else.

The stadium project critical path shifted from finalizing loads, stresses,

The construction of the Kibbie Dome changed the face of the campus—no doubt about that! But it changed and enlarged our spirit too. We could do big things and dream big dreams.
—Senator James A. and Louise McClure

Aesthetically pleasing patterns of webs clipped to billets above and below in the 7½-foot-deep plenum area of the TRUSDEK. 35,000 galvanized webs were used ranging from 2-inch 16-gauge steel to 2¼-inch 13-gauge steel, depending upon stress areas. Photograph courtesy of Cline, Smull, Hamill & Quintieri Associates.

Where else can you

Roll up the Football Field and put it away until next season?

Hold a Track Meet on the world's largest indoor track?

Play eight Tennis Matches or nine Basketball Games at once?

And also hold a Trade Show or County Fair with 'no rain' guaranteed?

Have rock concerts or band concerts to crowds of 21,000 or more?

—from a publication by Cline, Smull, Hamil Associates, Architects

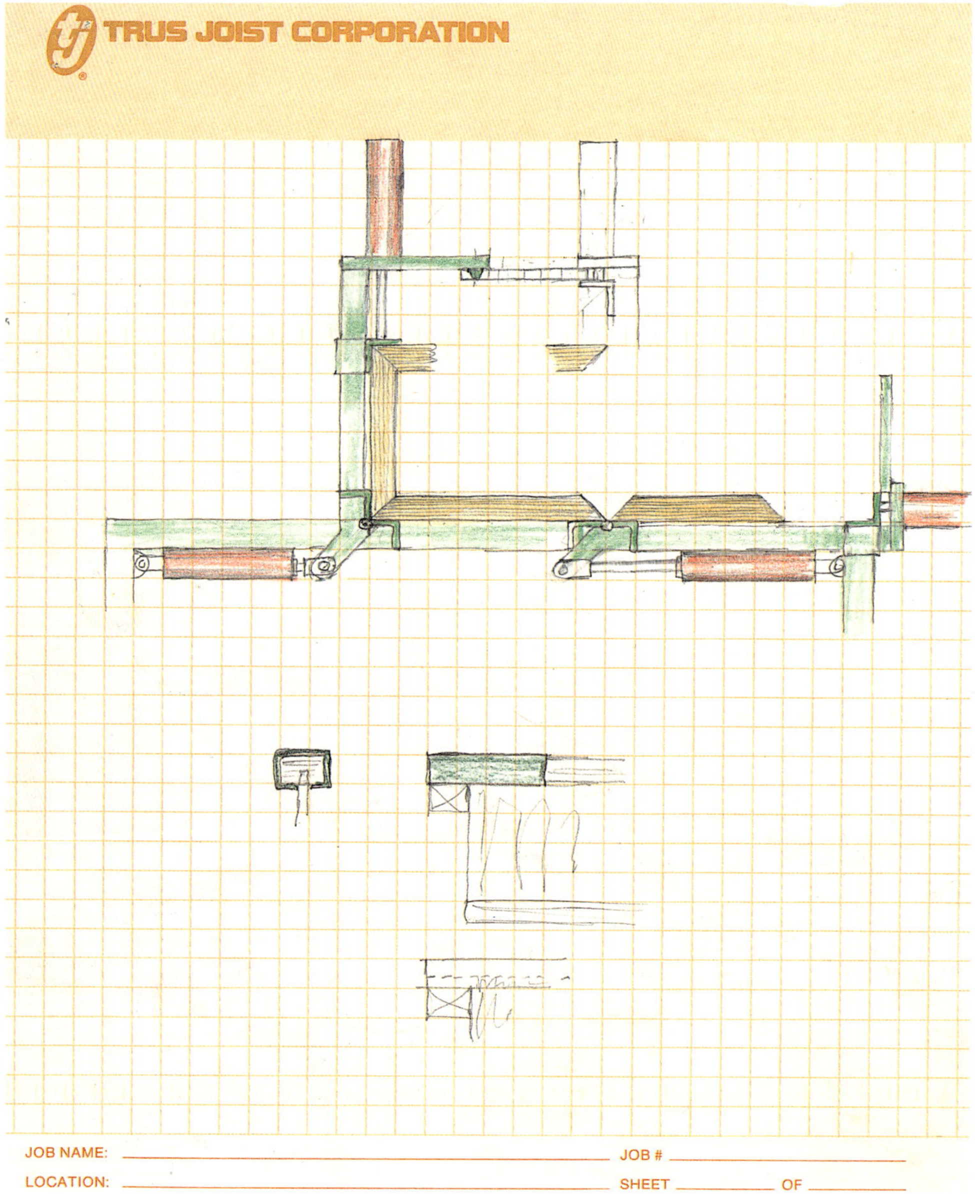

Hand drawing by Art Troutner conceptualizing the assembly and manufacture of a box beam.

design, and material specifications; to the production and delivery of the billets from Eugene; and finally to the assembly and erection of the barrel arch at Moscow. No slack—no idle time—existed on the critical path. All activities had to be successfully accomplished the first time they were attempted, if the company were to avoid embarrassment and penalties. There would be no "overs."

Zimri Mills led the design of the full-scale tests of TRUSDEK sections. Six tests conducted in early 1975 used load cells to mechanically pull cables attached to the ends of 30-foot billets, which were mocked-up with top and bottom chords, webs, and clip connections to be used in the TRUSDEK configuration. Tested "in compression to destruction" to verify the performances of the MICRO=LAM® chord members, these mock-ups showed that the chords failed in a coordinated "sine" mode. They buckled longitudinally between the panel points, which were in line across the several billets in the test sample, and then shattered. This mandated staggering the billet/web/pin connections on each side of the 2-foot-wide billets.

Another full-scale test was performed in the middle of January with

Precision-drilled, aluminum sealed, numbered, weather-protected, and precisely ordered billet packages ready for loading on special rail cars for shipment to Moscow, Idaho.

Jerry York of KKBNA and key Trus Joist personnel present. The staggered panel points proved successful, allowing for the completion of the final design and material specifications.

When combined with other analyses and tests, the full-scale tests confirmed the 1⅞-inch thickness of the MICRO=LAM® billets (down from the originally estimated 2 inches), as well as the panel layout, the related arch depth, the web diameters and gauges, and even the clip design that connected webs to billets.

Troutner insisted on capitalizing on the inherent strengths of all materials as well as assuring that they could be efficiently manufactured and assembled in the field. For example, KKBNA had specified a ¼-inch thickness for the 4,000 steel splice plate connections that would connect the billets. After testing and analyses, Art Troutner pointed out that a thickness of ⅛ inch met the design criteria with a generous allowance for safety. Troutner's experiences as an inventor had made him suspicious of engineering that specified over-designed structures "just to be safe." After hearing Troutner's reasoning, KKBNA agreed to reduce the thickness of the plates.

This was not the only instance where KKBNA made adjustments after a challenge by Art Troutner. When Troutner reviewed KKBNA's calculations of wind loads in the design of the arch support trusses that would connect the barrel arch to the pins along the bearing walls, he found that KKBNA's numbers had a safety factor twice what was required. Art confronted KKBNA and they again made appropriate adjustments.

"This was a stressful time," Mike Thompson said. "The work required for the final design of the barrel arch and the design and specifications for all the components was overwhelming. Every individual piece of the arch had to be produced accurately within extremely tight tolerances. That was a must if we were to zipper together thousands of these parts and connect them at the pins along the bearing walls within plus or minus 1/16th of an inch."

Shortly before manufacturing was to begin on the TRUSDEK billets, Sherm Nelson reported jokingly to me that KKBNA's Jerry York had made slight of the tight tolerances Trus Joist was specifying for wood.

"You'll have thousands of saw kerfs, each of them greater than the tolerances you are specifying," York had said. "What makes you think you will be able to maintain your dimensions within tolerances? We're discounting your ability to do it."

KKBNA, as the engineer-of-record, had the authority to determine and were professionally accountable for the final design and all analyses upon which they and others relied. If the stadium were to fail and the cause of that failure could be traced to KKBNA's work, then they would be liable for damages. Despite KKBNA's excellent reputation, this off-hand statement stuck in my mind. Several days later I went to Sherm Nelson.

"Sherm," I said. "That comment by KKBNA was not professional. It suggests a less rigorous attitude than we want and expect. My confidence has been eroded and I am uncomfortable. I'd like you to run a check on virtually everything they have done, computer runs and all. I won't ask you to apply your professional seal to their work but I want you to do the next best thing.

"Push everything else aside for a few days. Lock yourself in your office and don't come out until you would be comfortable placing your seal on everything KKBNA has done. I won't be at ease until you complete this exercise."

Sherm Nelson had made it clear that all his other responsibilities did not give him time to act as the engineer-of-record, KKBNA's role, for the trussed barrel arch. Yet he hesitated only briefly before accepting the assignment. He sequestered himself out of reach of people and phones and went to work.

Sherm and I had a good relationship that had solidified the year before during a six-hour walk into the early hours of the morning in Dubuque, Iowa, where Trus Joist had a manufacturing plant. The discussion had been over what role a professional engineer should play within Trus Joist and concerned aspects such as marketing, manufacturing, and research and development. We had struck an understanding that in exchange for an engineer's commitment to be creative and flexible in the exercise of his professional expertise, the company would respect the engineer's professional judgment as final.

I expressed it this way: "The company recognizes that engineers are bright and exceptional people who have gone to great effort and expense to acquire knowledge vital to the safety of others—knowledge about which those others are unaware. To use that knowledge creatively and prudently in the design of

One of 35,000 bolted clip, web, billet connections. Troutner's clever design with offset bolts allowed the clip to swivel and accommodate any billet expansion and contraction.

Trus Joist products, engineers must be free to make judgments that will be final and binding on other company personnel."

This understanding had substantially healed antagonisms between engineers, managers, and salesmen, and set more clearly which decisions would be made by each.

Ten days later Nelson emerged and reported his findings.

"I found four errors—two through miscalculation and two dimensional."

Either of the dimensional errors would have resulted in the structure not coming together at the pin connections along the bearing wall. Repairs could not have been made, and work would have had to stop until new materials could be manufactured and delivered.

Nelson advised Jerry York of what he had found, and York corrected the work to conform with Nelson's findings just days before the 2,300 billets were to be manufactured in Eugene, Oregon.

Chief Engineer Zimri Mills was a stickler for quality on all Trus Joist products. With the help of Art Troutner and others, Mills had developed the quality control procedures used in manufacturing, packaging, and shipping, and similar procedures would be used on the TRUSDEK barrel arch. In fact, it would be through quality control procedures that the company assured not only the intrinsic qualities of the various materials but also the maintenance of strict dimensions and tolerances, often in thousandths of an inch.

Mills insisted that an approved plan showing side and end view dimensions and quality control procedures be in place, in addition to the standard quality assurance provisions of Section 4.0 of the Trus Joist Engineer's Manual, before any component of the final barrel arch was manufactured.

This exercise in care didn't end with the manufacture of the components. Copies of the quality control procedures to be used during assembly and erection were furnished to the MacGregor Triangle Company crews well before work began on the site. Among other instructions, these procedures emphasized the critical importance of protecting the billets from moisture. Furthermore, Mills prepared for the University of Idaho a comprehensive set of procedures for maintenance of the system over time, including a requirement that the University control humidity in the plenum area with several large fans.

> *The Dome is a Godsend for my program . . . since its opening it has become the Mecca of the Northwest for meets and practices.*
> *—Michael W. Keller, University of Idaho Head Track Coach*

Final Design to Reality

BY 1974 MICRO=LAM® laminated veneer lumber technology had been advanced by Art Troutner and his research and development staff to a confidence level that promised substantial rewards. Immediately, the company made the major decision to commercialize the technology. MICRO=LAM® would soon compete favorably with solid structural lumber, steel, and concrete.

One of our newer managers, Bill Cooley, had shown his effectiveness working the bugs out of the new and advanced MICRO=LAM® continuous presses at the company's first manufacturing facility in Eugene. Cooley held an architecture degree, but his career path had taken him into management in the plywood industry. There he had acquired a vast working knowledge of veneer, adhesives, kiln drying, lay-up, and quality control. This know-how proved to be invaluable during the remaining MICRO=LAM® development and demonstration work, after Trus Joist had made its huge commitment in Eugene.

This knowledge, combined with Bill's organizational and people skills, made him a natural to accept the challenge of manufacturing the 2,300, 80-foot long billets to tolerances unheard-of in structural wood. When integrated into the arch's design as top and bottom chords, this application of MICRO=LAM® would cover the largest existing indoor college facility in the nation.

"The project came to the Eugene plant at a perfect time for us," Cooley said. "Being early winter we weren't particularly busy. So we had the space and manufacturing capability to respond. Additionally, this sort of activity would be exciting and challenging. Everyone at the plant signed on, which was really positive in our ability to respond to the unique complexities of the project.

"Everything we did was on a first-time basis. We felt the pressure of knowing that every billet we manufactured and hole we drilled had to work perfectly. Our staff came together in developing the sophisticated procedures for the manufacture, storing, conditioning, cutting, drilling, sealing, packaging, and

Erection begins in designated working areas off the ends of the playing field with the assembly of sections 80 feet long, 12 feet 5 inches wide, and 7½ feet deep on jigs curved to the arch configuration.

> *The preparation, performance and visibility of Idaho athletic programs have been tremendously enhanced by the ASUI Kibbie Dome. This facility is an important element as we recruit student-athletes, coaches, and members of the student body. It permits us to attract and stage events, some of which would otherwise not be possible. These events have and will continue to have a very positive effect on the University of Idaho and the Moscow business community. The "dome" is truly a multi-purpose facility, and has come to symbolize the home of the Idaho Vandal Athletic Program.*
>
> *—Kathy Clark, Senior Associate Athletic Director*

shipping of the 172,800 lineal feet of billets. We numbered and packaged each billet so that the erection people in Moscow could identify the material in precise order for field assembly and erection.

"After about half of the billets were manufactured, but luckily not drilled, I received instructions to cut them all back to precisely 24 inches so that both edges of a billet could be relied upon to form an accurate surface from which to measure and drill holes. That's the way things were—it was an advanced research and development project. There were times I felt I was in a one-man rubber raft in the middle of the Columbia River's turbulent current. The harder I paddled the faster I seemed to turn in circles."

Fortunately, Bill Cooley had the help he needed. He credited Keith Erickson, Eugene's Quality Control Supervisor, who was soon to become Assistant Plant Manager, with rendering a vitally important tool for this phase of the project. "Erickson authored what was a virtual step-by-step procedural manual that became the Bible of manufacturing and was meticulously followed by the Eugene manufacturing team. It worked perfectly."

In yet another action to ensure quality, on February 7 Cooley received from Jim Lyons a comprehensive set of drawings and a material list for a dimension check assembly (i.e., a partial assembly of the TRUSDEK system so the critical dimensions could be checked for accuracy) to be performed at the Eugene MICRO=LAM® plant. Mike Thompson shipped the prescribed webs, clips, pins, bolts, and splice connectors from Boise. Then he and Joe Smith flew to Eugene with two employees of MacGregor Triangle Company to help assemble the prototype structure, consisting of about one-quarter of a full arch. Once it was assembled, they, along with the quality control personnel at the Eugene plant, conducted and recorded strength, weld, dimension, and tolerance checks on a regular schedule around the clock. They found everything a go.

Bob Kunesh, Trus Joist's Wood Technologist, had been working the issue of moisture's effects on the billets. The problem had scuttled that first meeting after bids were opened when Kunesh announced that if unprotected, the billets could swell up to a cumulative total of 24 feet along the length of the building.

Kunesh had pushed all else aside and immersed himself in the problem.

He drew from an experiment he had conducted at the University of Michigan years before. "We were checking the way moisture came into wood in a tangential, radial, and longitudinal direction," he remembered. "To establish a control to measure moisture absorption, we applied aluminum paint to the two dimensions that were not being tested. The aluminum had sealed the surface and held the moisture out. I drew from that experience. We needed a solution in a hell of a hurry!

"I knew from the lab tests and my experience with wood that it was the tangential absorption (causing expansion across a billet's width) that I had to worry about. We had to be assured that the 1⅞th-inch billet edges were sealed.

The next step in construction was to connect three 80-foot sections on more elaborate scaffolding to form a half arch, after which the precision manufactured arch-support steel trusses were connected.

The first half arch, weighing 45,000 pounds, is lifted high above the playing field by a 125-ton crane equipped with 190-foot booms. MacGregor engineers designed large steel lifting trusses, which were connected to the half arch to distribute stresses evenly.

Next in importance was to prevent longitudinal absorption by applying aluminum paint across the ends of all billets to seal the end grain."

I convened the Steering Committee to hear Bob's recommendations. Although the cost had not been in the bid, orders were soon placed for the materials Bob recommended, including 400 gallons of expensive aluminum paint to be shipped to Eugene, Oregon, in 50 gallon barrels. Bob's specifications called first for one coat of vinyl acrylic emulsion sealer/filler diluted 2:1 on the edges and ends of all billets. This less expensive preliminary seal reduced the amount of aluminum paint required.

Once the primer had dried, two coats of leaf-type aluminum paint were applied to the edges and ends of every 80-foot-long billet. Bob prepared detailed written quality control instructions to ensure that the billet bundles would be protected in transit and at the job site with a watershedding cover.

"Peter called me at 3 A.M. the morning after we had made the decision to use aluminum paint," Bob said. "He asked whether I was convinced that this would solve the problem. I told him that I felt good about it and to go back to sleep."

Bob Kunesh wasn't alone working on the billet swelling challenge. Concerned about expansion due to moisture, Art Troutner came up as usual with his own imaginative solution. While not compromising the need to meet design stresses, he redesigned the billet clips. Inserting bolts through holes drilled in the opposite corners of each clip would allow it to swivel in the narrow space between billets and accommodate expansion and contraction.

Jerry York's concern that Trus Joist would be unable to maintain the tight tolerances further complicated matters. Bill Cooley recalled, "To alleviate his fears we constructed two special chambers at the Eugene MICRO=LAM® plant—one to preheat all billets for not less than 24 hours prior to drilling, and the other to maintain a separate 70-degrees Fahrenheit controlled space where each billet was drilled in specially designed and rigid steel jigs, complete with brass bushings. You'd think we were making a Swiss watch."

Dry and warm. That's what the Kibbie Dome has come to mean to me since its completion. I remember those cold, wet, blustery fall days back in the 1960s when we bundled up, bunched together, and shivered our way through football games in Neale Stadium. Now it's short sleeves and 70 degrees . . . the Kibbie Dome . . . warm and dry.

—Larry E. Craig, United States Senator

To help MacGregor Triangle assemble the chords and other structural materials, billets were packaged with all top chords together and all bottom chords together. In addition, the twelve shorter billets that completed the bottom of each arch were coded and then packaged in sequence to facilitate assembly.

After the billets were manufactured, drilled, and sealed with aluminum paint, they were carefully wrapped with water-resistant paper. Then they were loaded on custom-fitted, extra-length railroad cars for transport to a railsiding in Moscow, Idaho.

Too long for conventional trucking equipment, the 80-foot packages were transferred to a tractor with a special detached trailer operated by a teamster who, wrestling a large horizontal steering wheel, guided the end of the trailer around the difficult turns to the job site. The first of 20 railroad-car loads of completed billets left Eugene on Monday March 3, 1975, ahead of schedule.

To protect them from water and other damage at the job site, billet packages were supported off the ground on 6-inch by 6-inch wood stickers spaced close enough together to support the billet packages above the ground and prevent any noticeable sag. During erection, whenever it rained, exposed truss chords were covered. Additional aluminum paint was supplied to MacGregor to be used in the event of damage to billet edges or ends.

As cranes held half arches in place, arch support steel trusses were connected by large, greased pins to the steel supports encased in reinforced concrete along the bearing wall above and behind the stadium seating.

Pinned connections between the arch support steel trusses and the steel supports embedded along the side bearing walls. These supports, located by two laser transits, had to be accurate plus or minus $\frac{1}{16}$th of an inch, pin to pin and across the 400-foot clear span.

These preventative measures proved to be prescient. During transport from Eugene some of the paper billet wrapping was torn away and rain frequently pelted the job site, as it does often during the spring on the Palouse. On one occasion a foot of snow had to be shoveled off the stacks of billets. But that wasn't the worst.

Bob Kunesh reported to me, "On an inspection trip to Moscow I arrived to find the end of an arch assembly immersed in rain water in a hole that had been dug to accommodate assembly. Thank goodness we had the end seal on. I measured the billets. There had been no swelling."

Perhaps our simplest task was manufacturing the more than 35,000 cold rolled, zinc galvanized steel webs which, depending upon stress areas, ranged from 2-inch, 16-gauge steel to 2¼-inch, 13-gauge steel. Punching webs from tubing on large hydraulic presses had long been a standard function in the production of all the company's open-web trusses. To further guarantee tight tolerances, Joe Smith prepared an elaborate jig, which secured the flattened and punched end of a web on a pin before the other end of the web was punched.

> *I was the Student Body President at the time the decision was made to complete the roof of the Kibbie Dome. . . . I sensed it would be a unique contribution to the University and to Moscow. In discussing it with many members of the student body, they shared that vision.*
> *—Dirk Kempthorne, United States Senator*

Joe Smith was the manager of the new products plant where early models of Art Troutner's ideas were built, including prototypes of the new machines to manufacture the new products. Joe was the right man for the job. He knew how to craft Art Troutner's ideas into wood and steel and to make almost any "Rube Goldberg" scheme work. Smith more than made up in experience, intelligence, and common sense what he may have lacked in formal education.

True to the pledge he had made to me, Troutner moved quickly to finalize the clip design that would connect the webs firmly to the billets. He settled on two welded steel angles with holes punched for the bolt and pin connections.

Determined to keep costs down, Joe Smith experimented with various devices before settling on a simple jig that would hold the materials firmly in place while they were welded. To ensure accuracy within tolerances of .005 inch, there could be no movement during the fabrication of the 35,000 clips.

Joe Smith invented a new tool which ensured that the 2-inch by 5/8-inch pin inserted through two connecting webs and a clip would remain secure both during assembly and once the TRUSDEK was in place. This tool hydraulically swedged—that is, swelled—both ends of the pin without diminishing the web's movement around it.

A portentous event occurred with the design and construction of the arch support steel trusses that connected the barrel arch to the pins on the bearing walls behind the bleachers. There had been a misunderstanding as to which party would complete the drawings for these critical components, get them ordered, and have them arrive at the job in time for erection.

"It was a blustery, snowy day in January when we drove up to Spokane to the shops of Artistic Iron to check on progress," Thompson said. "We were stunned to learn that the drawings for these large trusses hadn't been completed, nor had the steel for their fabrication been ordered. This discovery blew our critical path. We had a genuine crisis on our hands.

"Fortunately our design-construct team pulled together and worked the problem out on a very fast track. KKBNA completed the drawings for the trusses and emphasized the critical importance of maintaining the specified tolerances. We stressed this point to Artistic Iron and assured ourselves that they were capa-

Scene of the work in progress with packaged billets in the foreground, assembled arch sections immediately behind, then half arches on scaffolding, and, finally, an early portion of the erected barrel arch spanning 400 feet and high enough so a twelve-story building could fit below it!

ble of performing. The crisis was averted and the trusses were completed, on schedule and within budget."

The pin connections to be embedded in the concrete along the side bearing walls behind the bleachers had to be accurate plus or minus 1/16th-inch, pin to pin and across the 400-foot clear span. This required the precision of two laser transits to get a fix on each pin's location.

Nothing was left to chance. Knowing that there would be movement within the pin connections after erection, KKBNA prescribed that each pin be greased during installation to prevent offensive squeaks later.

Although assembly and erection of the barrel arch was the responsibility of the MacGregor Triangle Company, Joe Smith spent about a month on the job in Moscow to assist them. He remembered, "I drove into the yard and thought: 'What kind of g— d— idiots are we?' The building looked like it was going to be a half-mile across! We put the webs, clips, and pins together in 80-foot lengths on a big table. Once these were assembled, twenty workmen picked them up and walked them across the playing field to the billets, which were secured in the required contour configuration."

Mobile 125-ton crane lifting the half arch for transport to the finished barrel arch progressing in the background.

Joe Smith told me, "Looked like a big snake going out through there." The clips were then attached to the billets with 5⁄8-inch bolts and torqued with air guns.

One reason Smith was called to the job was to help figure out how these thousands of connections could be made accurately and efficiently. This work was actually part of the erection subcontract with MacGregor Triangle Company, but the task of preassembly had stumped their job engineer.

When the decision was made to stagger clip locations across billets to achieve optimum design efficiency, the riddle of how to preassemble pins to webs to clips was born. Had the staggering not been necessary, the assembly would have been straightforward and simple.

"At first I thought the preassembly would be easy, even with the staggered clip locations," said Joe Smith. "I had prepared some sketches in Boise and sent them up with the materials, but they weren't working. I went to the job and began diagramming different combinations. I worked a week on the riddle, sometimes spending all night.

"That engineer of MacGregor's came in about then and started telling me that Trus Joist was going to be penalized if I didn't get this thing figured out. I got real mad and told him I was doing his job for him and he could have it right back if he wanted it. That settled him down."

Bill Schulte, Trus Joist's engineer on the job with whom Smith had been working, announced that he was going to build a small model to work with. Smith told him: "I have put so much time into this I can't stop to build a model. I'm going to keep working with paper and pencil in my motel all night if necessary to figure this thing out.

"Through a process of elimination, I finally found an assembly that fit. It was just in time—I was about to tell Art he had to modify something, but that would have cost precious time and money. I used to like puzzles, but I didn't after

The Kibbie Dome is my "home away from home." Since the Dome was constructed, I have had the opportunity to work over 1,200 games of one kind or another in its friendly confines and have enjoyed every minute of it knowing full well the weather would be great—always a constant. The Dome offers spacious broadcasting facilities that are as good as you could find anywhere in America, with the privacy of enclosed booths and great acoustics that would make any voice sound good.

Some great athletes have participated in sports in the Dome including Marvin Washington of the New York Giants; Mark Schlereth of the Denver Broncos; John Friesz, a Walter Payton Award winner now with the Seattle Seahawks; and Doug Nussmeier, another Walter Payton Award winner who is now with the New Orleans Saints. And how could anyone forget that the Dome once housed the world's best athlete, Olympic and world record holder in the decathlon, Dan O'Brien, who has made his own tributes to the Kibbie Dome.

—Robert R. Curtis, "Voice of the Vandals"

Workman tethered for safety tightens bolts at clip connections as he "zippers" together another half arch to the 455 lineal feet of barrel arch already in place. Note the guy wires securing the assembly to anchors on the ground. Photograph by Phil Schofield.

The arch's graceful geometry of web patterns and billet spacing catch the sunlight. The crane in the background is dwarfed as the stadium cover progresses toward completion.

The University of Idaho is fortunate to have the Kibbie Center Complex. . . . The facility is remarkable, not only for its structural innovations and functional versatility, but for how it continues to foster a sense of cooperation among many individuals and groups that benefit from its existence.

However, in spite of this two-decade long pattern of shared use, we are beginning to experience limitations in the building's ability to balance all the competing demands. We now are planning for phased additions to the Kibbie's west end to further enhance its capabilities to support the university and the greater community and respond to new challenges and opportunities in athletics, recreation, and special events. We are optimistic that this impressive complex can continue to support the ever-increasing variety of needs we house here.

—Robert A. Hoover, President of the University of Idaho

that." This was another demonstration of Smith's remarkable ingenuity. After that the preassembly work proceeded without incident.

Mike Thompson told me Bill Schulte was an aeronautical engineer who had worked at Boeing on the 747. He was an active member of the John Birch Society and had threatened to leave the company when he learned that Chairman Harold Thomas was looking into selling products to Russia. "Bill was a highly competent and practical engineer who had just passed his structural exams for Idaho," Thompson recalled. "Despite his strong political beliefs, he quickly became pivotal as an engineering conscience on the project.

"Bill was just the engineer we needed. He could quickly analyze a problem, make a decision, and go to work. Once Bill and I were standing boot-deep in mud when MacGregor's superintendent asked about the flex limits of MICRO=LAM®. His workers were assembling sections of arch and he needed an immediate answer. Bill pulled out his hand calculator, entered a series of numbers, and gave the answer. Bill's capability, good judgment, and decisiveness kept the project moving forward time and again."

The erection began with the assembly of packages roughly 80 feet long, 12 feet 5 inches wide, and 7½ feet deep on jigs curved to the arch configuration. This took place in designated working areas off the ends of the playing field. These assemblies were then transported to a more elaborate scaffolding where three of them were connected to form a half arch, and anchor hardware was attached to one end. It took some time for the bugs to be worked out of this process.

"I recall the first day Joe Smith was on the job," Thompson said. "MacGregor's superintendent had their staging area all set up. Joe became more agitated as he walked around and observed what he considered the fumbling and slow pace of the assembly process. MacGregor's superintendent, Allan Johnson, was just as stubborn as Joe and I could see their relationship deteriorating fast. Both were about to explode.

"I grabbed Joe and we walked out into the muddy field and found Allan Johnson. Joe began by asserting that Allan's program was going down the toilet fast unless some things were changed. Allan defended himself. The tension was palpable. They shouted at one another for several minutes, which released some

Aerial view of the project showing staging area and partially completed barrel arch. Note the herringbone pattern of the ¼-inch plywood sheathing being applied over the finished barrel arch to protect billets from moisture.

of their tension. An hour later their communication had improved and with it their relationship, mutual respect, and the assembly process."

Alternative erection schemes were evaluated and the chosen procedure emerged. One half of an arch at a time would be lifted into its final place. Two mobile 125-ton cranes equipped with 190-foot booms would pick up a 45,000-pound half arch with large steel lifting trusses specially designed by MacGregor engineers to distribute stresses.

The cranes would walk the arch like a great kite down the playing field, where it would be connected at three pins on the bearing wall. A portable tower crane would then receive the half arch and hold it in place until the two mobile units returned with the other half arch and its three pins were installed on the opposite bearing wall.

After that, the three cranes would gradually lower each half arch fractions of an inch at a time until they meshed at the apex one hundred and fifty feet above the playing field. Finally, prepunched ⅛th-inch steel splice plates would be bolted in place to secure the two half arches. An unnerving wind blew the morning erection was to begin. Bets in local saloons were running heavily that the first arch would not come together at the top. When the bettors learned that this connection had a tolerance of only ⅛th inch, the odds against success increased.

The wind blew harder. Glen Cline was there when the cranes maneuvered the first two half arches into position and workmen installed heavy pins at the bearing connections. Once they were secured, courageous workmen scrambled up the twelve-foot wide arches to insert the final bolts through the splice plates and tighten the nuts, thus completing the first arch successfully and without incident.

"They came together just at sunset," Glen said. "It was a beautiful picture with the sun coming through. There was clapping, a few hoorays. Mostly people just stood and looked at it, awestruck."

Once the first arch was erected, connected, and safely guyed, it was learned that the bolts could not be inserted into one clip connection at the top. It had turned dark by then so the problem was left for the night.

"The next morning I was up early and climbed back up," Joe Smith

The campus-enhancing, award-winning Kibbie-ASUI Activity Center Dome under construction in 1975.

Vertical, insulated end wall preassembled on the ground using open web TJMs® and TJHs® is lowered into place by crane. A clever connection between the fixed end wall and the flexing arch accommodates movement laterally and vertically.

Photograph courtesy of Cline, Smull, Hamill & Quintieri Associates.

said. "I found a note tacked to a billet that read: 'If you want a good time tonight call room 1102 in Theophilus Tower.' Some damn kids had crawled up there during the night. That same morning Art was there and came out on the job. I told him about the connection problem and he almost fainted."

"That can't be," Troutner said. "It can't do that."

"Come on up and I'll show it to you," Smith responded.

"Well, if you say so I'll believe it."

Art Troutner wasn't about to climb up that narrow arch. He had served in World War II as an aerial engineer in B-17 and B-29 bombers, and had vowed never to fly again. Fortunately for everyone, "When we put the next arch up and connected it, the first straightened up and the bolts went right in," Smith recalled.

The next day the next two half arches were set into place without incident. As workmen inserted bolts from above and tightened nuts from beneath, the clips at each panel point "zippered" together the new arch assembly along the 455 feet of standing arch. On June 11, 1975, only 24 working days after the first arch was lifted into place, the final section of the cover was secured to the foundation and its adjoining arch.

Smith mused, "Once we solved all the glitches everything went so smooth it was almost unbelievable."

Only one near disaster occurred during erection. Smith remembered, "It was the stationary crane that almost caused the catastrophe. The operator began to drop his one half arch. The inside billet hung up on the ones already installed and the top billet warped up badly."

Joe Smith saw this and spoke to the crane operator, but the operator ignored him.

Fully aware of an approaching calamity, Smith walked over to Allan Johnson, MacGregor's superintendent, and spoke with an earnest authority. "If you don't want to take that half arch back down and repair it then you'd better get that operator to do what I asked."

A certain disaster was averted.

There were also some humorous incidents. Art Troutner had concluded that all of the narrow spaces between butting billets should be filled with epoxy. Smith purchased the best material he could buy and climbed up on the arch.

The Kibbie Dome was a subject of disbelief in the planning stages. A covered football field in Moscow, Idaho, was widely regarded as a flight into fancy. But the reality has proved to be a blessing for the University and the community. The Dome has been far more than a home for our Vandals. At Commencement it has protected the parents who come to celebrate the graduation of offspring. It has made cultural events, such as the acclaimed Jazz Festival, possible. Trade shows have made commercial use of it with economic benefit for local businesses. Our students in ever-increasing numbers go to the Dome in their commitment to physical fitness. Our life here would be poorer without it.

—Malcolm M. Renfrew, Emeritus Professor of Chemistry and Graduate of the University of Idaho

With generous room for fans, the Vandals take on Boise State in a basketball game in the spacious Kibbie Dome. The playing surface can be changed to accommodate football, track, and other athletic events.

"Art had a thing with epoxy so I tried to squeeze it into the narrow cracks between the splice plates," Smith said. "The epoxy wouldn't cooperate and dropped in big globs the size of chestnuts on one of the new cranes and the epoxy started to smoke. I took it and threw it out. We decided instead to fabricate some thin metal shims to fill the narrow gaps. When these were inserted between the billets and the last bolt was loosened, the billets came together nicely and all the bolts fit perfectly."

It rained heavily while erection was moving forward and the exposed top chords of the arch began to cup, rising as much as one inch in the middle of a 24-inch-wide billet.

Everyone became alarmed, particularly KKBNA. They announced that this condition had not been anticipated in their calculation and might change the structural integrity of the MICRO=LAM®. They insisted that an independent wood engineer inspect the warping and assess the possible damage.

Mike Thompson recalled, "While the engineer was inspecting the billets, it stopped raining. Before he could issue his report, which concluded that there had been no damage, the billets dried out and the cupping disappeared." Between storms, workers scrambled like ants over the four acres of billets to secure the plywood roof sheath that would cover the dome. "We took advantage of the good weather by accelerating the diagonal placement of the ¼-inch plywood sheathing across the entire barrel arch."

Earlier in 1975, after about one half of the barrel arch had been erected, KKBNA insisted on testing another model of the arch in a wind tunnel located in the Midwest to confirm the dome would hold up under unbalanced snow and wind loads. While the top and bottom chords of the arch were designed to be in compression, unbalanced loading might result in portions of the TRUSDEK being placed in tension. Jerry York of KKBNA became concerned that a combination of extraordinary wind and snow loads might occur simultaneously. Arrangements were quickly made to test a mock-up with winds of 100 miles per hour with a ten-foot-deep snow load on the windward side, clearly a worst-case scenario.

Following this test, KKBNA mandated that as a counter force several thousand sand bags, each weighing about fifty pounds, be placed on the leeward

The University of Idaho Kibbie Dome is not only a great venue for football, it is an arena where I enjoy watching basketball, track and field, and even an occasional rodeo!
—John Friesz, Seattle Seahawks quarterback

Campus culture has been enhanced with such activities as the Lionel Hampton Jazz Festival and this half-time band performance.

side of the barrel arch at strategic panel points along the lower chord of the plenum area.

"When we placed this huge and urgent order for sand bags in the middle of the winter at the Pioneer Tent & Awning Company in Boise, they thought we were crazy," Thompson said. "But they met their deadline and the quality was excellent."

Art Troutner thought the sandbags were overkill, of course. It didn't surprise him when, ten years later, they were removed after the performance of the barrel arch proved that the structure did not require such counterweights.

The last wind tunnel test had also indicated something else. Because the arch was not perfectly rigid, it would flex under varying loads. In addition to moving laterally, the top of the barrel arch could move up-and-down and side-to-side. Up until then, only the lateral movement had been contemplated. Now the connections between the arch and the vertical trusses that framed the end-walls of the stadium had to be redesigned in a hurry. Trus Joist rushed to design, and KKBNA approved, a clever connection between the fixed end-wall and the flexing arch, which would accommodate movement in two planes. The barrel arch had to be free to move laterally and vertically, and the stationary end-wall had to be firmly connected to it. The solution was a connection which permitted slip to occur in the vertical and horizontal direction while providing firm anchorage for the wall perpendicular to its surface. This was accomplished in a similar manner common to highway bridges where one of the supports is permitted to move, either by slipping on a teflon surface or by using a rocker device. The end-wall was then freestanding from the base receiving support at the top only when required by wind forces. Trus Joist's standard TJM® and TJH® open-web series trusses were used as the structural members of the end-walls.

The remaining work, mostly by subcontractors, proceeded on schedule. The entire project was completed and ready for the University of Idaho's first home football game that year. When it was dedicated on October 11, 1975, the structure was officially named the William H. Kibbie-ASUI Activity Center Dome in honor of the Associated Students of the University of Idaho and William H. Kibbie, a former student of the University and generous contributor to the project.

I go back a few years, and I remember playing and then watching football games and track meets in rain, wind, snow, clouds, and darkness in Neale Stadium. Now being able to watch athletics in the comfort of this magnificent facility is truly a benefit to the fans as well as the participants. The Kibbie Dome has been a great resource for students and the public extending far beyond varsity sports. We are fortunate to have such a beautiful and useful arena on campus. Who could have imagined at the time of construction that the Kibbie Dome would come to be cited as one of the three earliest indoor football fields in the world? We are proud of the people who made it possible.

—Flip Kleffner, Retired Director, Alumni Relations

The Dome serves the community in many ways by bringing indoors circuses and shows like this boat and mobile home display. Even an international model airplane competition was held in the stadium because of the remarkably calm air.

Essences and Honors

FALL, WINTER, AND spring bring severe weather to Idaho's Panhandle, but the new and unique structure, with a seating capacity of 18,000, walled out the menace of weather and opened the University of Idaho to year-round programs bringing a renewal that continues to this day. Football, basketball, tennis, and track, as well as many nonathletic campus and community programs, take place inside away from the rain, snow, and cold.

It was the high point in Trus Joist's history; it's never been exceeded.
—Arthur Troutner, Genius Inventor

Renewal has come in many forms, including an improved home game winning record. An uninterrupted, year-round flow of activities has changed the University's profile, scope, and culture forever. Student and public attendance at events has multiplied several fold, from hundreds into thousands. In February the University's annual Lionel Hampton Jazz Festival alone attracts over 12,000 fans and participants from the United States and Canada.

Skeptics of the project had proclaimed it an imprudent risk, but the key players drew heavily upon their imaginations, skills, and self-confidence to unflinchingly perform crucial roles in an audacious act of creation. As Art Troutner reflected, "It was the high point of Trus Joist's history. It's never been exceeded."

Mike Thompson had this to say: "The successful completion of the TRUSDEK barrel arch over Moscow's ASUI/Kibbie Dome was then the largest single event in Trus Joist's history. It shifted Trus Joist from a good solid company marketing proprietary structural products to a world-class organization. It was one of those pivotal events in a company's history that changes a company's image of itself and how others perceive it."

Many recognized the achievement as a masterpiece where talent, technology, vision, and commitment were manifest in a unique creation, and Trus Joist's barrel arch stadium cover received wide recognition. *Engineering News Record, Architectural Record, Forest Products Journal, Western Building Design*, and other major magazines carried extensive coverage about all phases of construction of the building. The concept drew interest from overseas, as evidenced by articles about the Dome published in Japanese and German.

The biggest surprise came in 1976 when the highly esteemed American

Blending with the rolling hills of the Palouse and walling out the menace of harsh weather, the William H. Kibbie-ASUI Activity Center Dome accommodates many events and forms a major component of life on the University of Idaho campus. Photograph courtesy of Cline, Smull, Hamill & Quintieri Associates.

Society of Civil Engineers (ASCE) honored the project. In April, Dr. Ernest W. Hartung, President of the University of Idaho, received the following letter from Eugene Zwoyer, Executive Director of the ASCE: "It was a great pleasure to wire you that the Barrel Arch Roof of the University of Idaho Stadium has been selected to receive the ASCE 'Outstanding Structural Engineering Achievement Award' for 1976. My hearty congratulations on behalf of the Board of Directions and the 72,000 members of this Society."

This prestigious award is bestowed annually upon the one project completed in the previous year that stands above all others. The runner-up that year was a remarkable multi-billion-dollar pet project of then Governor Nelson Rockefeller, the New York State South Mall in Albany, New York. That the judges selected for this award a stadium cover that cost less than one million dollars testifies to its uniqueness. Thus, the stadium cover on the campus of the University of Idaho joined other esteemed engineering creations awarded the honor, including the St. Lawrence Seaway, the World Trade Center, the St. Louis Peace Arch, and the John F. Kennedy International Airport.

President Hartung wrote in a letter to me, ". . . on behalf of the University, I would like to extend to you and your colleagues in Trus Joist a word of deep appreciation. The facility is working out beautifully and in many ways is exceeding our fondest hopes for it. It is a rare and brilliant design which you and your colleagues have executed to perfection, and we at the University are all deeply appreciative. The ASCE recognition honors all who had a part in the design and execution of the project. Please extend a personal word of congratulations to Mr. Troutner."

The raising of the roof over the University's athletic field had raised the standards by which wood-construction architecture is judged. When the University orchestra struck up the first graduation march the next year, it harmonized with the three-dimensional music of the new stadium cover.

The facility is working out beautifully and in many ways is exceeding our fondest hopes for it. It is a rare and brilliant design which you and your colleagues have executed to perfection.

—Dr. Ernest W. Hartung, President of the University of Idaho (1965–77)

For Further Reading See:

The University of Idaho Stadium: A MICRO=LAM® Application, by Sherman A. Nelson, P. E. (F. ASCE), September 1976.

University of Idaho Stadium, A MICRO=LAM® Laminated Veneer Lumber Application, by Sherman A. Nelson, P. E., September 1982.

"Creative Engineering in Structural Wood," *Western Building Design*, September 1975, pp. 11-13.

AFTERWORD

BEAUTY BENEATH THE SKIN BY BILL BOWLER

The campus of the University of Idaho in Moscow, Idaho, boasts a number of beautiful architectural creations including the handsome brick and stone Administration Building, with its venerable clock tower built in 1909. The building that dominates the skyline today is the contextual Kibbie Dome, a structure that mirrors the rolling Palouse and serves to wall out the coarse weather and shelter students, faculty, alumni, and friends who come to the campus for a range of activities.

The Kibbie Dome's functional beauty manifests below its skin in a structure that is undeniably a great work of genius and beauty. There a remarkable 400-foot clear span wooden-trussed barrel arch unfolds—the first of its kind—which even today remains unduplicated in the world of architecture.

This magnificent structure, masked by a cover which displeases some, serves as an unwitting metaphor for Idaho. Too often Idaho's culture, if the state can be said to have an identifiable, homogeneous culture, possesses an undercurrent of brilliance and ingenuity that others fail to see because of isolated examples of extremism and what is perceived to be provincialism.

Those of us who live in Idaho, though, see a much different picture. We see a wide scattering of small, handsome, cohesive communities. We see individual effort, determination, and inventiveness as hallmarks of our existence in this grand western experiment. Idaho has been, and will become even more, a state of monumental achievements in the sciences, in engineering, in entrepreneurship, in preserving ecological harmonies, and in the arts.

The daring and genius of Idaho are truly exemplified in the architecture and design of Art Troutner, Glen Cline, and the vision of Trus Joist's executives.

They turned the impossible into the obvious. They overcame monumental risks in bringing the quintessential Kibbie Dome to the University of Idaho. We must hold steadfast to this characteristic that has distinguished Idaho since its beginning—our willingness to assume the chance of loss so that something truly great can be born.

The Kibbie Dome towers as proof this quality prevails today. When Trus Joist took on the contract to build the stadium cover, the factory and equipment to manufacture the primary component, MICRO=LAM®, was not yet available, nor was the technology to produce it fully developed. The executives at Trus Joist bet the fledgling company's future on the strength of a brilliant design and a faith in the ability of the team assembled to commercialize the technology.

The completed Kibbie Dome is a magnificent tribute to the vision of this design team and the determination it took to see the project through to fruition. This same vision and venturesome spirit are at work today renewing Idaho's culture—not one based on frontier rusticity or on individualism gone awry—but one that is rooted in a farsighted civic and ecological consciousness and directed toward stable, sustainable industrial, commercial, cultural, and civic growth.

And Idaho is maturing on another level as well—philanthropy. Because of the philanthropy of William Kibbie, the final funds were assembled which cleared the way for completion of the stadium cover. More recently, committed civic leaders together with wealthy citizens have created foundations in Idaho with generous endowments. This expansion of corporate and individual philanthropy signals both the flowering of culture and the opportunity to further nurture the bloom through our own brand of ingenuity and flair.

The Kibbie Dome stands as testimony to our capability and proof that through skillful and intelligent design, fearless resolve, and dedicated determination, we can lead the country into the next millennium as an example of what the American dream really means.

TECHNICAL TERMS AND DEFINITIONS

PROVIDED BY SHERMAN NELSON

Arch: A structural member, in this case a curved shape, that requires its support to be both vertical and horizontal.

Barrel arch: An arch as a continuous member in a third dimension so that it appears to be a portion of a barrel form.

Billet: The raw form of the laminated veneer lumber as it exits the press. Essentially it is simply a plank which can later be cut up or modified into the specifically required application for the material.

Buckling: The tendency for a structural member under compression to deflect from its axis—as with a wooden yardstick stood on one end and pushed on at the other end causing it to buckle under the load and take on a deformed shape.

Chord or chord member: In a truss, one of the principal members—the top or bottom members of the truss braced by the web member of the truss.

Clear span: The distance between supports.

Clips: The name given to the steel connecting detail at the panel points.

Compression: To load a structural member along its axis in a manner that tends to shorten its size in that direction.

Jig: A template or guide made in order to assist in the proper handling or manufacture of a construction detail.

Laminated Veneer Lumber: A manufactured product of laminated veneers with the grains of the veneers all aligned in the same direction (as opposed to plywood, where the grains of the veneers are cross-oriented).

Load cells: In this case a hydraulic device used to impose a load on the test, as with a hydraulic jack.

Machine stress grade wood: Lumber specifically graded for structural uses.

MICRO=LAM®: The brand name of Trus Joist Corporation's laminated veneer lumber product.

Open web truss: A structural member, specifically a truss, where the configuration consists of top and bottom chord members at the extremity of the depth connected by web members, in the form of triangles.

Panel points: In a truss, the points of connection of web members to the chord members.

Panels: In a truss, the segment of a chord member between the panel points.

Plate connections: Steel plates with holes drilled therein through which bolts are inserted. The means used to connect the MICRO=LAM® billets end to end.

Plenum: The open space in a roof system used for access or for air flow or utilities.

Saw kerf: In this case, the result of a circular saw blade cutting into a wood board and the resulting void of wood.

TRUSDEK: An earlier proprietary structural system developed by Trus Joist Corporation similar to the system utilized in the stadium.

Truss: An assemblage of structural members forming a rigid framework.

Web or web members: In a truss, the members that are between the chord members, usually in a triangular configuration along the length of the truss.